The TR-3B Black Manta

The TR-3B, or Black Manta, or Astra is one of the most fascinating potentially top secret aircraft ever created by the United States military.

The big reason for this is that this craft is supposed to use anti-gravity technology. This technology is discussed a lot in this book.

Also, what other evidence is there for the existence of this craft? There are lots of unknowns here but there are also enough pictures of this aircraft that you can't just discount sightings as being something else which the observers misinterpreted.

There are even anti-gravity patents by US government researchers which might be applicable to the reality of the TR-3B.

Overall, the existence of this craft is a huge mystery which still needs exploration and solutions to what it really is.

This is what I'm trying to do in this book.

The TR-3B Black Manta

The TR-3B Black Manta

Copyright Page

This book is copyrighted for 2024

Title: The TR-3B Black Manta

Subtitle: What is it Really?

The Aliens and UFO Secrets Series Book 8

By Martin K. Ettington

All Rights Reserved USA 2024

ISBN: 9798303896048

Printed in the United States of America

The TR-3B Black Manta

The TR-3B Black Manta

Other books by Martin K. Ettington

Spiritual and Metaphysics Books:
Prophecy: A History and How to Guide
God Like Powers and Abilities
Enlightenment for Newbies
Removing Illusions to Find True Happiness
Using the Scientific Method to Study the Paranormal
A Compendium of Metaphysics and How to Guides (Six books together in one volume)
Love from the Heart
The Enlightenment Experience
Learn Your Soul's Purpose
Pursuing Enlightenment
A Modern Man's Search for Truth
Use Intuition and Prophecy to Improve Your Life
The Handbook of Spiritual and Energy Healing
Spiritual Masters

Longevity & Immortality:
Physical Immortality: A History and How to Guide
The Commentaries of Living Immortals
Records of Extremely Long Lived Persons
Enlightenment and Immortality
Longevity Improvements from Science
The 10 Principles of Personal Longevity
Telomeres & Longevity
The Diets and Lifestyles of the Worlds Oldest Peoples
The Longevity Six Books Bundle

Science Fiction:
The History of Science Fiction and Fantasy
Out of This Universe
Personal Freedom-Parts 1 & 2
The Psychic Soldier Series:
    Book 1-Himalayan Journey
    Book 2-A Soldier is Born
    Book 3-Fighting For Right
    Book 4-Earth Protector
The Immortality Sci Fi Bundle
The Immortals of the Interstellar Colony
Evolving Humanity

The God Like Powers Series:
Human Invisibility
Invulnerability and Shielding
Teleportation
Psychokinesis
Our Energy Body, Auras, and Thoughtforms
How Premonitions Really Work

The God Like Powers Series—
Volume 1 Compilation
The Reality of Ghosts & Spirits

The Yoga Discovery Series:
Yoga-An Ancient Art Form
Hatha Yoga-Helping you Live Better
Raja Yoga-Through the Ages
The Yoga Discovery Package

Business & Coaching Books:
Creating, Paublishing, & Marketing Practitioner Ebooks
Building a Successful Longevity Coaching Business
Why Become a Coach?
The Professional Coaching Success Trilogy
2020-Make Money Writing and Selling Books
The 2020 Handbook of High Paying Work Without a College Degree

Science, Technology, and Misc.
Future Predictions By and Engineer & Seer
The Unusual Science & Technology Bundle
Removing Limits On Our Consciousness-And Thinking Outside the Box
Strange but True Stories and Facts
The Microscopic World Inside and Around Us
Radionics and Life Force Technologies
Infinity and Our Unbounded Universe
Planet Earth is Conscious: And Life Exists in Amazing Places
Accepted Science Which Is Likely Wrong
All About Ball Lightning
Stranger Than Science Stories and Facts
Stranger Than Science Stories and Facts (Book Two)
Quantum Mechanics, Technology, Consciousness, and the Multiverse
Universal Holistic Philosophy

Legendary Animals and Creatures

The TR-3B Black Manta

Are Cryptozoological Animals Real or Imaginary?
Fire in History and Mythology
All About Dragons
Sea Serpents and Ocean Monsters
The Importance of Fire in History and Mythology
Thunderbirds: Legends and Reality

Ancient History
The Real Atlantis-In the Eye of the Sahara
Ancient & Prehistoric Civilizations
Ancient & Prehistoric Civilizations-Book Two
The History of Antediluvian Giants
The Antediluvian History of Earth
Ancient Underground Cities and Tunnels
Strange Objects Which Should Not Exist
More Out of Place Artifacts
Strange and Ancient Places in the USA
A Theory of Ancient Prehistory And Giant Aliens
The Big Book of Pyramids Around the World
Underwater Ruins of Civilization
A Theory of Ancient Prehistory and Giant Aliens

Aliens and Space
Aliens and Secret Technology
Aliens Are Already Among Us
Unidentified Submerged Objects and Underwater Bases
Four Evidences of Aliens and UFOs in Earth's History
Human And Alien UFO Anti-Gravity Research

Living in Space
Designing and Building Space Colonies
Humanity and the Universe
All About Moon Bases
All About Mars Journeys and Settlement
The Space and Aliens Six Books Bundle
The Space Colonies and Space Structures Coloring Book
All About Asteroids
Spaceships, Past, Present, and Future
Astronauts, Cosmonauts, and Other Important Space Flyers
All About Mars Journeys and Settlement
Mining the Asteroid Belt
Exploring and Settling Our Huge Solar System
James Web Space Telescope Mysteries

Survival
33 Incredible True Survival Stories
How to Survive Anything: From the Wilderness to Man Made Disasters
Building and Stocking a Nuclear Shelter for less than $10,000
Survival of Humanity Throughout the Ages

Time Travel
Real Time Travel Stories From a Psychic Engineer
The Real Nature of Time: An Analysis of Physics, Prophecy, and Time Travel Experiences
The Multiverse: Time and Dimensional Travel Q&As
Stories of Parallel Dimensions We Live in a Malleable Reality-and We Can Change It
Alternative Dimensions & the Otherworld
Quantum Mechanics, Technology, Consciousness, and the Multiverse

Self Improvement
The Importance of Genius in our World
Creating Your Own Reality
The Fear of Failure: And What You Can Do About It
Building Hope and Wonder Among Chaos.
A New Paradigm of Truth and Happiness
See The World Clearly: Be Happier and More Fulfilled
Stress Relief and Methods to Do So
The Importance of Creativity and How to Improve Yours Building Self Confidence

Political & Social
The Suppression of Truth in the United States and the World
Fifteen Ideas Which Will Change the World
The Empire of the United States: Forged by the Spirit of God in Man

The TR-3B Black Manta

The Longevity Training Series

(A transcription of the online Multimedia Longevity Coaching Training Program)
The Personal Longevity Training Series-Book1-Long Lived Persons
The Personal Longevity Training Series-Book2-Your Soul's Purpose
The Personal Longevity Training Series-Book3-Enable Your Life Urge
The Personal Longevity Training Series-Book4-Your Spiritual Connection
The Personal Longevity Training Series-Book5-Having Love in Your Heart
The Personal Longevity Training Series-Book6-Energy Body Health
The Personal Longevity Training Series-Book7-The Science of Longevity
The Personal Longevity Training Series-Book8-Physical Body Health
The Personal Longevity Training Series-Book9-Avoiding Accidents
The Personal Longevity Training Series-Book10-Implementing These Principles
The Personal Longevity Training Series-Books One Thru Ten

These books are all available in digital, printed, and audio formats from my website and on Amazon, Barnes & Noble, Apple ITunes, and many other sites.

The TR-3B Black Manta

If you have any questions about this book or other subjects please contact the Author at:

mke@mkettingtonbooks.com

My Books Website is:

http://mkettingtonbooks.com

The Online Training Course site based on these books is:

https://www.mkettingtonbooks.com/training-home

Signup for our Monthly Newsletter and download past issues here:

https://mobile.mkettingtonbooks.com/newsletter-signup-and-archive/

The TR-3B Black Manta

Table of Contents

1.0	Introduction	1
2.0	The Origins of the TR-3B	3
3.0	TR-3B Recent History	7
4.0	An Interview with Ed Fouche	10
5.0	Podkletnov Anti-Gravity	21
6.0	Navy Anti-Gravity Patents	31
7.0	TR-3B Design	35
8.0	Propulsion Technology	37
9.0	Aircraft Sightings	49
10.0	Actual Photographs	55
11.0	Claims of this Aircraft in Desert Storm	59
12.0	The Silent Triangle	69
13.0	Summary	73
14.0	Bibliography	75

The TR-3B Black Manta

The TR-3B Black Manta

1.0 Introduction

The TR-3B, or Black Manta, or Astra is one of the most fascinating potentially top secret aircraft ever created by the United States military.

The big reason for this is that this craft is supposed to use anti-gravity technology. This technology is discussed a lot in this book.

Also, what other evidence is there for the existence of this craft? There are lots of unknowns here but there are also enough pictures of this aircraft that you can't just discount sightings as being something else which the observers misinterpreted.

There are even anti-gravity patents by US government researchers which might be applicable to the reality of the TR-3B.

Overall, the existence of this craft is a huge mystery which still needs exploration and solutions to what it really is.

This is what I'm trying to do in this book.

The TR-3B Black Manta

The TR-3B Black Manta

2.0 The Origins of the TR-3B

The Secret TR-3B Government Spacecraft Capable of Interstellar Travel

The TR-3B is a secret government spacecraft that has been flying in our skies since 1994. You can do your own research on this craft, however because it is a classified project you may not find much on this. I have done lots of digging, and I think it is a strong possibility that this craft actually exists. It is indeed powered by an onboard nuclear reactor. It is completely silent and can be seen as a triangular shaped craft that has 4 lights. It has one light on each corner of the triangle along with a light that is in the center. All of these lights can be seen on the underside of the craft. This aircraft also has a code name of Astra. This aircraft has advanced stealth technology that is even far more advanced than the B-2 stealth bomber. It has a

The TR-3B Black Manta

polymer skin that allows it also to change its reflectiveness and even change shapes to the human eye.

There are many videos on YouTube that show sightings of this craft. The government of course denies its existence entirely, however there are so many sightings, and scientists that have worked on the project that have come forward, that it should be considered as fact that it exists by the American public. The only problem is no one does research about it and very few people know that it actually exists. That is why I am writing this article. It is also known that the stealth polymer skin can even change the shape to look like a flying "cylinder". So many people have also documented and recorded crafts of this description flying in the skies of the United States.

This craft on top of having advanced stealth, can also trick radar equipment into thinking that there are multiple aircraft in other locations to serve as a decoy, and further allow this ship to remain undetected and to setup a very clever diversion. A circular plasma filled accelerator ring called a magnetic field disrupter surrounds the rotatable crew compartment. This technology is far ahead of any technology previously known by any government on the planet. Was this craft reverse engineered from other crash landed alien ships/technology?

The TR-3B is a high altitude stealth aircraft. It has an indefinite time it can sustain its altitude. Once you get it high up there and going fast, it doesn't take much propulsion to maintain its altitude and trajectory. At the Groom Lake military base there are rumors that there is a new element that acts as a sort of catalyst to the plasma reaction. This causes the vehicles mass to be reduced by exactly 89%. The craft can go at Mach 9 speeds both vertically and/or horizontally. The performance the TR-3B

The TR-3B Black Manta

is only limited by the stress the human pilots can endure, which is a lot, considering that along with the 89% reduction in mass, the G force is also reduced by 89%.

The start of the creation of this craft can be traced back to the 1940s and 50s. Mostly because of Project Redlight. Project Plato, originally established as part of Project SIGN in 1954 its purpose was to establish diplomatic relations with extra-terrestrials. This project was deemed successful when mutually acceptable terms were agreed upon. What those terms exactly were is unclear. However many high ranking individuals in secret government positions have come forward speaking about this. What I am sure of is that these terms involved the exchange of technology from aliens, while we did NOT interfere with alien affairs on this planet. The aliens agreed to provide MAJI with the means and the technology as long as they didn't interfere with the alien's agenda. I am assuming this had to do with the abduction and experimentation on the human population. This project is being continued on a military base in New Mexico.

Project Snowbird established in 1954 was created for the purpose of building a flying saucer type craft for the public. This project was successful when a craft was built and flown in front of the press. This project was used to explain most of the UFO sightings that were occurring in that time. This project was also used to divert the public's attention from Project REDLIGHT.

The objective of project REDLIGHT was to fly recovered alien craft, however the project was postponed after the death of many United States Airforce pilots that were killed in trying to operate these crafts. All of these test flights occurred in Area 51 (Groom Lake). Some of these test

The TR-3B Black Manta

flights also occurred at a secret military base code named DREAMLAND.

Project REDLIGHT resumed in 1972 when the flights of some of these recovered craft were partially successful and accompanied by black helicopters and f-15 fighter jets.

I have done lots of research on this subject and I am positive that our government is hiding this from us. Many of the UFOs that are seen in our skies are actually reverse engineered alien crafts built by the United States Airforce.

The TR-3B Black Manta

3.0 TR-3B Recent History

The public history of the TR-3B goes back to the nineteen nineties when the sightings of it started.

One of the most well-known accounts is from 1990, when a source (who remains unnamed) claimed to have worked on the craft in a secretive military project known as the "Black Manta" program. This individual reportedly described the TR-3B as a triangular-shaped craft, often hovering or moving with great speed, performing maneuvers beyond the capability of conventional aircraft.

<u>Black Triangle sightings in the US saw a significant increase in 1997</u>

According to the NIDS report, they received hundreds of calls about Black Triangles when they opened their reporting hotline in 1999, each one describing similar large, silent, and often slow-moving aircraft.

In order to assess the real frequency of Black Triangle sightings in the United States, NIDS contacted the Mutual UFO Network (MUFON) and Larry Hatch, who maintained one of the largest and most comprehensive databases of UFO sightings at the time. These organizations were particularly valuable to NIDS investigators because they didn't open their hotline to easily collect reports until 1999, limiting their sample size in the years prior.

NIDS assembled more than 700 combined Black Triangle sightings into a single U.S. highway map, with U.S. Air Force installations also shown as blue, yellow, and green circles.

The TR-3B Black Manta

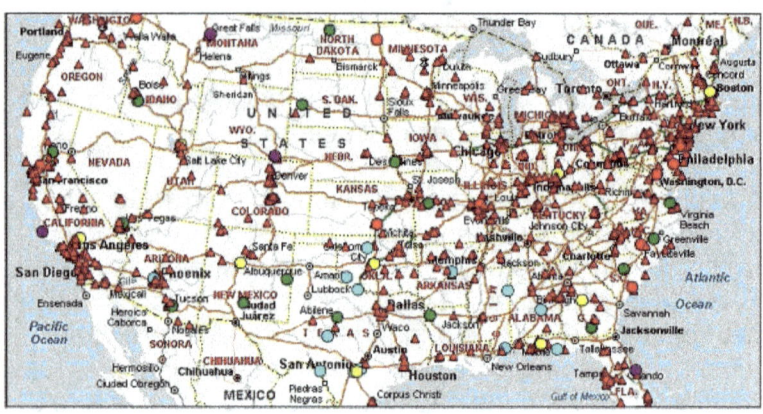

Triangular craft sightings are shown as red triangles, with circles denoting the location of U.S. Air Force installations (NIDS)

According to the NIDS report, all three databases saw a significant increase in black triangle UFO sightings starting in 1997. Further, the vast majority of sightings came from well-populated areas, which would be an unusual flight plan for a highly secretive government aircraft program.

Indeed, NIDS points out how the dispersion of these sightings is not in any way similar to sightings of the F-117 Nighthawk or B-2 Spirit during their testing regimes, which were largely limited to Nevada and Southern California and seen over sparsely populated areas.

The TR-3B Black Manta

NIDS vs MUFON vs Hatch

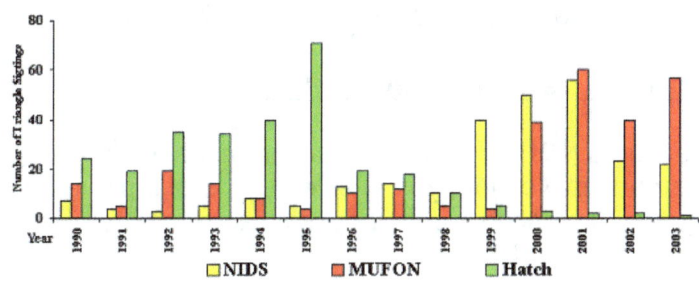

The frequency of Black Triangle reports by year, as collected by each database. (NIDS)

NIDS initially hypothesized that these black triangles were covert military aircraft, in keeping with the legend of the TR-3B, but the sheer volume of sightings over cities and near heavily trafficked interstates prompted NIDS to reconsider that idea, openly wondering if these objects may come from elsewhere.

Regardless of whether you believe in the TR-3B or Black Triangle UFOs, it seems evident that people were seeing *something* in the skies during this time

The TR-3B Black Manta

4.0 An Interview with Ed Fouche

Ed Fouche is well known for his books on UFOs and there are a lot of records showing he worked for the government on secret technology projects for many years. He also supposedly worked at Area 51.

Ed Fouche Telephone Interview 1999

Paola Harris (PH): During your talk at the Laughlin Conference in Nevada, you showed slides of a plane built using very advanced technology, similar to extraterrestrial technology.

Ed Fouche (EF): They show the development of certain craft which the government denies exist.

PH: Are the photos classified or unclassified?

EF: Some used to be classified, but they aren't anymore. Take the TR-3B for example. We had an old digital image where it was impossible to make out the shape of this craft. So I took it to a friend, a graphics expert, who enhanced the details by enlarging it. That's how we managed to accurately recreate its characteristics.

PH: Is this material in the public domain?

EF: No, we may have restored a photo, but the TR-3B is still the most secret craft in the world. The government has many covert programs, and at least half a dozen of them concern top secret planes— the most highly classified objects in the world—and the technology behind them.

PH: When triangular objects are sighted, as in Belgium in 1990, are they actually craft built by the government?

The TR-3B Black Manta

EF: They aren't alien. They have a flat, triangular shape; we call them *mantas* - a kind of TR-3a with bat wings, totally different from the TR-3B. A friend of mine, "Gerald", worked for **NASA** for the whole of his career. We met at the end of 1976. According to him, there are three prototypes of these craft measuring about sixty meters (200') across, and the operational model is about 180 meters (600') across. He saw them with his own eyes. The copyright on this information dates back to 1994.

PH: Did he tell you what year he saw them?

EF: I think he saw the first prototype in 1991. There were triangular planes before then, but this was the first triangular craft with a *Magnetic Field Disrupter* (**MFD**), a magnetic propulsion system. It's a plasma accelerator which somehow disrupts gravity around the craft, thereby reducing its mass and weight by 89%. In practice, this was the result of reverse engineering from UFOs, but it didn't fly like a UFO, as it had an operational efficiency of 89%.

If you've got a craft which can't reduce its mass and weight more than this, the only logical thing to do is to build a triangular-shaped craft with three multiphase rocket engines mounted at the corners, which is precisely what people see, both in the 200' version and in the 600' one. The photos, which are very darkmand are posted on the Internet, were taken at night, and you can see three bright lights which correspond to rocket engines. The large glow in the center is due to the energy generated by the MFD.

PH: A friend of mine who works in Intelligence told me that you had come into possession of *MJ-12 documents*. What did they contain?

The TR-3B Black Manta

EF: Part of the documents concerned the autopsy of some supposed aliens, similar to those described by Col. Corso in his statements, and also some things that came out of *Santilli's alien autopsy*, broadcast in the U.S. on FOX, which in my opinion is a reconstruction by the government in an attempt at disinformation.

PH: Are you referring to the Santilli footage with the alien with *six fingers?*

EF: Yes. The aim of the reconstruction, based on the original footage but showing an alien with six fingers, was to confuse the public. Anyway, an MJ-12 autopsy document also talks about a removable lens, like the one seen in the Santilli footage and as described by Col. Corso. But note that these MJ-12 documents *were filed a long time before Corso wrote the draft of his book or before Santilli had seen it in the footage.*

PH: That's interesting. And did the aliens have six fingers?

EF: No, four. Why did they film aliens with six fingers? Think about it for a moment. If everyone believes that four-fingered aliens are roaming the Earth, and all of a sudden footage appears which corresponds exactly to the real film but where the being has *six fingers*, then it's clear that the *footage is not genuine*. The aliens in the video aren't real.

PH: There may be more than one race, given that they are allegedly biological entities created in a laboratory.

EF: I disagree with Dr. Wolf's view here. Initially, there was a UFO crash in Germany and the aliens there were grays. The second crash happened at Roswell and another then occurred in the U.S. In between the two incidents, there was a UFO crash in Russia.

The TR-3B Black Manta

PH: Some alien races are believed to be very similar to human beings. Col. Corso said that it was worrisome that some of these races may be so humanlike that they could walk along Pentagon corridors without being recognized.

EF: As I mentioned, one of my key informers, an NSA investigator, swore that there was an alien race of this kind. We knew of their existence but had never managed to communicate with them. This is why a massive technological race began in order to equal them, so we could protect ourselves. Although these aliens have never shown hostile tendencies, they were so technologically advanced that they constituted a potential threat. You know how the military thinks: *if they can't control something, they prefer to kill it.*

PH: That's absolutely true!

EF: So it was all part of the rationale; it was a question of secrecy. At that time the *Strategic Defense Initiative* (**SDI**), or the so-called *Star Wars program*, was launched, and following a new Majestic 12 Charter, various technologies were developed, basically to defend the atmosphere, which forms our external barrier.

PH: Have you ever heard of luminous aliens, or beings of light?

EF: Quite the opposite. The reason why this race uses **EBEs** (*Extraterrestrial Biological Entities*), going back to Roswell, is because they are neither humanoid nor sentient beings. In the autopsy document, EBEs are described as absolutely identical to each other, that is to say, *they are manufactured beings*. This is logical, in my opinion. Indeed, there is no reason to send a living being

The TR-3B Black Manta

to explore the universe on a one-way journey when your scientific knowledge is advanced enough to create an EBE which can do it for you.

Today, we have moved closer to this type of technology:

Beings are being designed with higher perceptive capabilities, for example better visual abilities, where the bodies are programmed for a mission and then they are sent into space. While considering myself to be a rational person and not agreeing with many aspects of UFOlogy, everything I wrote in my book is completely logical, in my view and in the view of those who know me.

PH: So in other words, they are robots or androids. Do you know anyone who has communicated with them?

EF: According to Gerald, at the site of the Roswell incident a "body" was found still alive. Everyone denies that Secretary of Defense Forrestal was at the scene. On the contrary, it seems that everyone had a good false alibi.

Well according to Gerald, Forrestal was right there, and communicated telepathically with the alien. As you know, **Secretary Forrestal** then began to show signs of a mental and emotional disorder, and was admitted to Bethesda Hospital, where he threw himself out of a window on the sixteenth floor. In my view, there aren't any holes in the story.

They have never sent us an alien ambassador, and there are no underground bases where military personnel work alongside aliens. If we had had some sort of technology-exchange program with them, why would we have to invest billions of dollars in Research and Technology in order to do everything on our own? We could give them hundreds more human beings in exchange for their science.

PH: In your opinion, do these androids or manufactured beings have a program? They must have, mustn't they, seeing as too many people continue to talk about their abduction experiences?

EF: I'm not so sure, and this is part of the current diatribe. According to Gerald and another two people I have talked to, there have been some abductions, but only a limited number. *Most abductions are carried out by the government.*

PH: Why on Earth would the government be involved in abductions and insert screen memories in people's minds, making them believe they were abducted by aliens?

EF: There are several reasons. TMK is easy to use: it's a mind control technique first used in the '30s and '40s, when the **CIA** used to give people large doses of marijuana extract (THC), heroin and mescaline, altering human behavior via brainwaves or ultrasound to create screen memories or to put false memories into people's minds.

PH: In other words, you are saying that the manipulation of human beings is being carried out by the government, and not by aliens.

EF: Three people I trust swear that they are absolutely sure of this. Think about it—one of the reasons which comes to mind is that if there were some alien germ which could infect us, we could alter human genetics through injections. But what would happen if aliens came to Earth and released microbes which could modify our atmosphere? We would have be able to adapt. In other words, maybe we know that aliens are coming and they will settle on Earth. How can we protect our race?

PH: Is this how you explain the reasoning behind abductions?

EF: Exactly. It's for the protection of the species. Or maybe they've learned how to improve our species from genetic engineering.

PH: But why abduct people, Ed? Why do we have to do it in such a violent way? Can't we find another solution? For example, say it's being done to fight a disease?

EF: But how would you harvest eggs?

PH: But these people truly believe that they have passed through walls and been in spaceships. Why would the government create all these false memories?

EF: Disinformation. If you plant half a dozen different versions of what is going on in people's minds, no-one will ever believe anyone else, and that's exactly what's happening.

PH: It's a worldwide phenomenon. In Italy too, really strange things are happening to people who live both in cities and in remote areas. How can it possibly be a government conspiracy?

EF: It could be a cross-section.

PH: They would have to have made agreements with Italy and all the other nations.

The TR-3B Black Manta

EF: I have never believed in the *theory of aliens* or *the New World Order controlling minds* for financial reasons. The New World Order exists for a different reason, it's part of a *Majestic 12 plan* to unite the world under a single government. Supposing there's an alien threat, how could we unite the world under one power? How could we all agree? We would never agree, not even through the United Nations. Not unless a pre-existing authority imposed the terms of the future new order.

PH: So you believe that convincing people that we are threatened by aliens will unite everyone in a single world order?

EF: Yes. It would also happen in the event of a crisis. Y2K is a prime example. The world now relies on computers. There is no easier way than to crash all the computers in the world to take control and overcome those who possess all the computer-based military power.

PH: Why would they take control?

EF: To rule the world under a single government.

PH: Made up of...

EF: Whoever these people are.

PH: So, as far as you are concerned, the UFO phenomenon is in reality a major conspiracy devised by the people of the New World Order, who may be acting exclusively for their own benefit? We blame aliens for abductions, but in actual fact it's the government's doing?

EF: Put it like this: we are under threat. Whoever has sent these genetically designed androids or robots (EBEs) to

The TR-3B Black Manta

Earth has transmitted a return signal. These beings have found an inhabitable planet with intelligent life forms, and sooner or later *the real "striking force," the real aliens, will arrive*. So we have embarked on this massive program; we have eggs from all these people. We have created special, genetically modified children, and we have placed them in all areas of society through adoption, including placements with individuals connected to the government.

PH: Just a minute, you may be right! They have given these children to people connected to the government. I think I have met some of these children. But why would they do this?

EF: OK, I'll tell you. A microbiologist friend of mine is a great believer in *junk DNA;* that is to say, the potential of our DNA has never really been grasped. In reality, it can form a third DNA helix, which is an important evolutionary change. Once *the One World Order* is established, all the genetically modified individuals will be injected with another DNA element.

Imagine, for example, DNA capable of sparking off an evolutionary process inside them: these individuals will be able to control the world and not be influenced by aliens, even biological entities which communicate mentally. They will all be placed in key positions, and obviously their seed and eggs will be used to make more human beings, who will also be endowed with these abilities.

PH: Have you ever as a child had anything strange happen?

EF: When I was small, my mother says once when we were riding in the forest, we saw a forest fire which came right into the car but my mother does not like to talk about

The TR-3B Black Manta

it. I have a vague recollection of it. It is funny that you are the only person who has ever asked me that question.

PH: I am curious sometimes why people are interested in this field, Ed. Sometimes it could be they are chosen or given an invitation. Thanks for the interview.

The TR-3B Black Manta

5.0 Podkletnov Anti-Gravity

Dr. Podkletnov has been well known on the internet for almost 30 years due to the article about his rotating disk anti-gravity experiments. I remember when I was at Boeing that links to his articles were banned because so many employees spent time online reading them.

Eugene Podkletnov has a New Gravity Modification Experiment. Here is an interview with him which is reprinted:

Thirty years ago, Dr. Eugene Podkletnov developed a gravitational shield using high-speed rotating superconductors. Now he's testing a new device that he claims will generate better results — and we've got the video he says proves it. We join him to learn more about the details of his experimental claims and learn how his latest design may produce a gravitational field capable of lifting heavy objects…

Eugene, thank you for joining us. You've got a new experimental video out that's raising a lot of big questions, so I want to touch on that — but let's start

The TR-3B Black Manta

with what you're currently up to. Are you still working at the university in Tampere?

Yes. My current job is at Tampere University of Technology in Finland, teaching chemistry and material science. My work in the field of gravity research is a side project, something I've been doing for over 30 years now.

Dr. Eugene Podkletnov, research scientist at Tampere University of Technology

Some of that work I do here in Finland, some in Moscow, and I and recently completed some research in Czech Republic in Prague. I also have colleagues that I collaborate with in Italy, Canada, and the United Kingdom — such as Dr. Giovanni Modanese.

He is an outstanding scientist, and an excellent theoretical physicist, which has led to a great working relationship. I do experimental work for him and he does theoretical analysis for me, which is a good thing.

I want to touch on the new experimental video, which appears to be an experiment involving high speed disk? Is this similar to your original experiments rotating YBCO superconductors at high speeds, which purportedly created what was described as a gravitational shielding effect?

Yes. I'm continuing my work in this direction — but I'm not calling this effect gravitational shielding, but instead a *modification of the local gravitational field*. This comes from the work I began this work 30 years ago — learning how to use high-speed rotating objects with superconducting components to modify gravity.

The TR-3B Black Manta

What I've found, however, is that the superconductors are needed only to create a certain density of electrons, so in the experiments you seen in this video we're working with very thin gold layers which generate the same effect at room temperature.

My latest research shows that working with composite materials that do not include superconductors at all, we're able to create gravity fields, in vacuum, in the air, and so far in every object placed within the vicinity of this experimental gravity generator. It's a much more efficient method.

An ion-implantation machine similar to the kind used by Podkletnov

Okay, originally you were working with superconductors, but you now believe that the electron density is what made the effect happen, and you've replaced the superconductor with a thin gold film on normal conductors? Using this design, how much force are you generating?

The TR-3B Black Manta

Yes, that's right, and without using superconductors, we are now creating these effects at normal room temperature. There is no need for cooling.

Well, I can only share some estimates. Currently, I think we can generate a lifting force of about 300 to 500 kilograms per square meter. I've tested these devices in vacuum conditions and they appear to generate the same force, which rules out atmospheric effects.

It also appears that this experiment can generate either a repulsive force or an attractive one, depending on the geometrical configuration of the experiment. That's why we're not using the term "gravity shield" anymore, it just isn't appropriate.

We're simply trying to explain the experimental results that we got, and based on what we've been able to achieve, we're describing it as a modification of the local gravitational field.

Do you think this would this work in space, for instance, where there's no reaction mass to push against?

Well if it works in a vacuum, then perhaps it will work for space propulsion — but still you should understand that these are mechanical models, which use pretty heavy discs that are rotating at high speed.

For space travel, it would be better to try and find a way to create the effect using rotating magnetic fields rather than mechanical components, which is something I am considering.

The TR-3B Black Manta

A solid state device that isn't limited by mechanical rotation can be scaled much more easily — maybe even to thousands of kilos per square meter, which would be a remarkable result. If that can be developed, it would definitely be suitable for space propulsion.

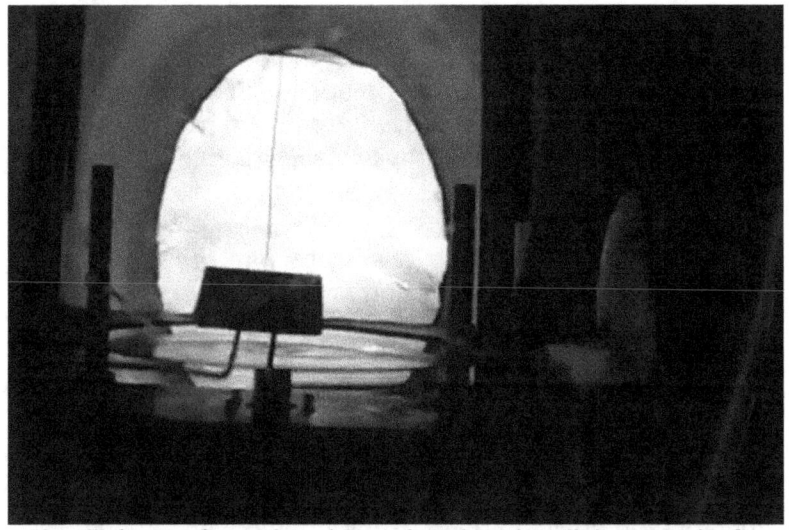

A still-frame from the video showing the disk, sample & retaining string in the vacuum chamber.

Now in the experimental video, it appears that there's an object suspended over the rotating disc on an armature that allows it to move up & down, and in one video you appear to have a restraining string to prevent it the object from scraping the rotating disk surface. Can you describe the actual experimental setup a little bit for me?

So we have a vacuum chamber. We have a disc which has special nano-coatings made with ion-implantation. The disc is rotating at this speed from 8,000 rotations per minute up to 12,000 rotations per minute, and at these speeds we

The TR-3B Black Manta

have a lifting force that affects all of the samples we've tested, regardless of composition. We've tried using glass, plastic, metal, even vapor — they're all repelled from the surface of the disc.

The force we're seeing produces the lifting effect, and seems to encompass the space around the sample with what you might call an envelope of the effect. These forces are vector fields, so they can be applied in any direction in space. I've hypothesized that what may be happening is that the disc is creating a gravity well, which the sample is then falling into regardless of orientation.

From what you're saying, it sounds like the thin film ion-impregnated gold coating is what may be producing this effect then? Can you elaborate a bit more on that for me?

Yes. The nano-coating of the disk consists a thin layer of gold, from 5 to 30 atoms thick, which is applied to the surface of an aluminum disk using a high-power ion-implantation device. I have experimented with various shapes for the coating as well as doping the coating with other elements, and settled on one consisting of concentric circles. I have experimented with doping the coating with other elements as well.

This is an embarrassing question because I'm sure you've taken it into account, but have you been able to firmly rule out experimental errors such as perhaps vibrations in the disk, the armature holding the sample, or maybe even an inductive effect that could be moving the armature?

Yes, it's a typical question that people ask. The samples that we use typically weigh from 30 to 50 grams, and using only vibration in the vacuum chamber, it's impossible to move these objects up to five to seven centimeters from the surface of the disc. So no, it's definitely not the vibration. When we begin rotating the disk, at any speed lower than 8,000 rotations per minute, we see no effect or movement. Then, as we reach 8,000 rpm, we see the effect begin to appear. So this effect only appears at certain speed of rotation.

This experiment can be easily reproduced at any laboratory, so researchers who have a serious interest they can ask me for help and I will explain for them how to recreate it.

Speaking of replications, in terms of the failed NASA replication for the rotating-superconductor experiment, I'd heard from a NASA insider that they only spun it up to about 200rpm. Does that mean that they didn't fully complete your experiment?

I cringe every time I hear that NASA failed to replicate my experiment, because no, they didn't fail. They made their own disks, and they were big enough: about 12 inches in diameter. They published some initial test information indicating that they had definitely noticed some unusual effects.

Then I got involved in participating to helping them to replicate my experiments, and they practically had everything ready when they ran out of money. So at the last stage they were not able to rotate the superconductor in the magnetic field, and shortly after that the department of defense came in and grabbed all the experiments. All of this research was transferred to Dr. Ning Li — so now NASA has nothing, and we have nothing either.

The TR-3B Black Manta

YBCO levitation is easily explained, but Podkletnov claims superconductors can also modify gravity

From what I've read, the YBCO superconductors were hard to construct & often self-destructed at high speeds. Given the simpler design & construction of (https://medium.com/predict/eugene-podkletnovs-new-gravity-modification-experimental-video-b7813b04c6f8, 2022)***this new device, does this mean the end for your superconductor research?***

It's important to understand that superconductors were extremely useful as model materials because they allowed us to create any configuration of the magnetic field that we wanted to — but being able to work without them using composite materials seems to give us practically the same effect or an even better effect than superconductors.

For the time being, we have come back to rotating disks, but we're using a composite structure now, and under certain conditions we're creating this gravitational modification. I also plan to begin working with rotating magnet fields, which are practically the same thing as a rotating body. But so, but in a, in a bit different way. So the mechanism is a bit different.

Let me close by asking whether you've noticed any other anomalous phenomenon other than the repulsive force you're demonstrating in the video clips?

Well, of course there are different phenomena that we can observe during our experiments. But frankly speaking, my focus at the moment is not in the scientific realm of categorizing anomalous effects — I am interested in the engineering work of maximizing the main propulsive effect to a usable level.

The reason for this is simple: the opportunities that this technology presents may greatly benefit humanity, and I would love to see this work develop towards a practical implementation for propulsion, perhaps to the point where we could use it to begin our travel to space and "boldly go" …you know the rest.

Dr. Eugene Podkletnov has a doctorate in materials science from Tampere University of Technology in Finland, and is a graduate of the University of Chemical Technology, Mendeleyev Institute in Moscow.

He has nearly two decades of experience at the Institute for High Temperatures in the Russian Academy of Sciences, and has authored a number of papers on experimental research on EM/gravitational coupling in superconductive materials.

The TR-3B Black Manta

The TR-3B Black Manta

6.0 Navy Anti-Gravity Patents

There have been literally hundreds of anti-gravity patents over the years. Of course we don't know how many really work.

The US Navy has patented antigravity. Or as described by the inventors: inertial mass reduction device. Could these be the same patents for technology being used in the TR-3B?

Patent No. : US 10,144,532 B2
Date of Patent : Dec. 4, 2018

The patent describes a device that uses a microwave emitter to create a high-frequency electromagnetic wave through a cavity to create a polarized vacuum. This polarized vacuum, in turn, reduces the mass of the vehicle containing the device.

A description of what this patent says can modify gravity follows:

It is possible to reduce the inertial mass and hence the gravitational mass, of a system/object in motion, by an abrupt perturbation of the non-linear background of local space-time (the local vacuum energy state), equivalent to an accelerated excursion far from thermodynamic equilibrium (analogous with symmetry-breaking induced by abrupt changes of state/phase transitions).

The physical mechanism which drives this diminution in inertial mass is based on the negative pressure (hence repulsive gravity) exhibited by the polarized local vacuum energy state (local vacuum polarization being achieved by a coupling of accelerated high frequency vibration with

The TR-3B Black Manta

accelerated high frequency axial rotation of an electrically charged system/object) in the close proximity of the system/object in question. In other words, inertial mass reduction can be achieved via manipulation of quantum Held fluctuations in the local vacuum energy state, in the immediate proximity of the object/system.

Therefore it is possible to reduce a crafts inertia, that is, its resistance to motion/acceleration by polarizing the vacuum in the close proximity of the moving craft.

Polarization of the local vacuum is analogous to manipulation/modification of the local space time topological lattice energy density. As a result, extreme speeds can be achieved.

If we can engineer the structure of the local quantum Vacuum state, we can engineer the fabric of our reality at the most fundamental level (thus affecting a physical systems inertial and gravitational properties). This realization would greatly advance the fields of aerospace propulsion and power generation.

It seems that the inventors propose to use the properties of high energy electromagnetic field generator to interact strongly with the vacuum energy state. The system described seems, according to the authors, to generate quantum fields' fluctuations permeating the entire fabric of space-time. Which in this case, the aim is to alter the material to create an antigravity field...

The TR-3B Black Manta

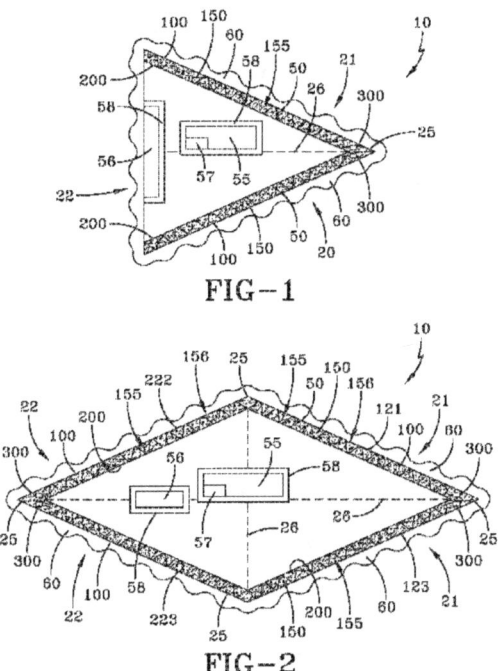

FIG-1

FIG-2

It seems that the inventors propose to use the properties of high energy electromagnetic field generator to interact strongly with the vacuum energy state. The system described seems, according to the authors, to generate quantum fields' fluctuations permeating the entire fabric of space-time. Which in this case, the aim is to alter the material to create an antigravity field…

The TR-3B Black Manta

The TR-3B Black Manta

7.0 TR-3B Design

All of the pictures and diagrams and pictures of this aircraft show it as a triangle with propulsion emitters in the corners of this triangle. Three multi-mode thruster engines provide the actual energies to propel this craft.

The skin has various materials to make the craft stealthy too.

There is also supposedly a highly compressed mercury plasma in a circle with magnetic confinement which is accelerated thousands of times per second to provide a significant decrease in apparent gravity. The entire craft is also nuclear powered.

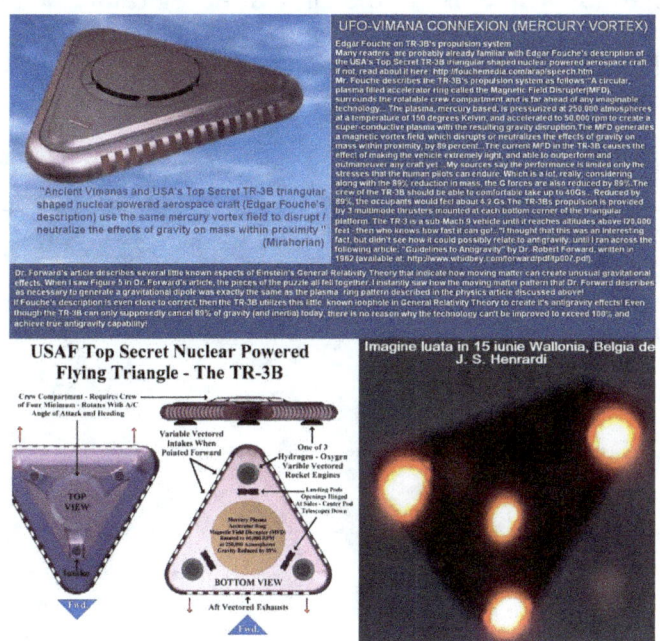

The TR-3B Black Manta

A circular plasma-filled accelerator ring called the Magnetic Field Disrupter [MFD] surrounds the rotatable crew compartment and is far ahead of any imaginable technology. Sandia and Livermore National laboratories developed the reverse-engineered MFD technology. The mercury-based plasma is pressurized at 250,000 atmospheres at a temperature of 150 degrees Kelvin, and accelerated to 50,000 rpm to create a super-conductive charged plasma with resulting gravity-disruption [reduction of almost all of the pull of gravity and effects of inertia].

The MFD generates a magnetic-vortex field which disrupts or neutralizes the effects of gravity by 89 percent on a mass within proximity. The MFD creates a disruption of the Earth's gravitational field upon the mass within the circular accelerator. The mass of the circular accelerator and all mass within the accelerator, such as the crew capsule, avionics, MFD systems, fuels, crew environmental systems, and the nuclear reactor, are reduced by 89%.

The current MFD in the TR-3B craft causes the effect of making the vehicle extremely light, and able to outperform and outmaneuver any craft yet constructed - except of course those back-engineered total-antigravity craft, which the government does not admit exist.

The TR-3B Black Manta

8.0 Propulsion Technology

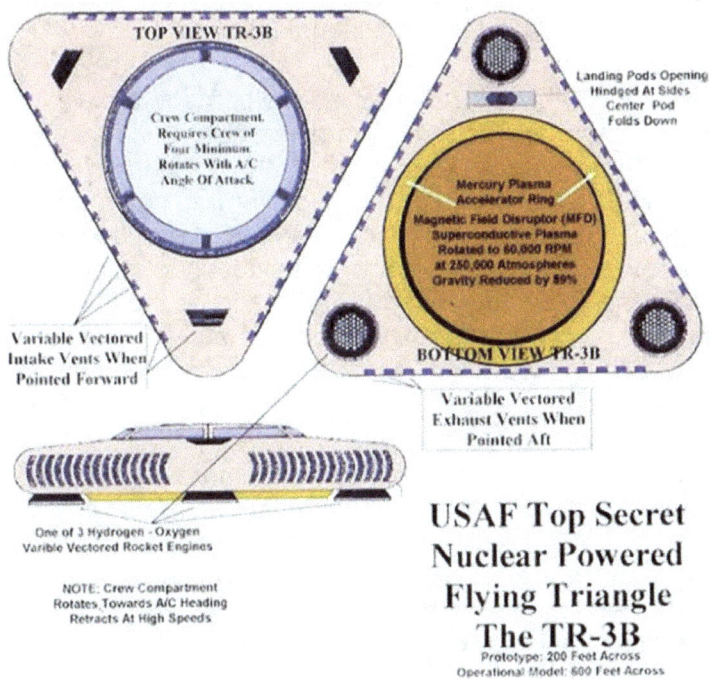

A) First description of anti-gravity technology in the TR-3B

The TR-3A may have been a fairly grounded story about a stealthy reconnaissance platform designed to operate alongside the F-117... but this story gets a whole lot stranger when discussing longstanding claims of the *TR-3B*. Unlike the turbojet-powered TR-3A, the TR-3B is supposedly powered by a reverse-engineered anti-gravity drive that was recovered from a crashed alien spacecraft. The TR-3B is where reports of UAP, performing seemingly impossible aerial maneuvers, intersect with stories about very real aircraft.

The TR-3B Black Manta

There are lots of claims all over the internet about the TR-3B's anti-gravity drive, most of which include using nuclear power to rotate highly pressurized mercury to produce plasma, and in turn, a gravitational field.

Most stories about the TR-3B's development revolve around Operation Paperclip – the program that saw the United States absorb more than 1,500 German scientists and engineers at the end of World War II to continue work on various defense technologies. We've discussed in the past how myths about advanced Nazi technologies were born in the years after World War II, and as the folklore surrounding this anti-gravity platform matured over the years, it was not exempt from the reach of "Wunderwaffe" claims.

Despite how silly these claims sound, there is actually *some* evidence to substantiate elements of them. For example, an anti-gravity drive of the sort believers claim powers the TR-3B would require massive amounts of power production – more than we could possibly produce with current aviation engines. But, then, Lockheed Martin *does* have a patent for containerized cold fusion reactors they say could be small enough to fit inside the fuselage of an F-16…

B) Second description of this technology by Ed Fouche

Another description of this propulsion system is by Ed Fouche. Mr. Fouche describes the TR-3B's propulsion system as follows:

"A circular, plasma filled accelerator ring called the Magnetic Field Disrupter, surrounds the rotatable crew compartment and is far ahead of any imaginable technology…

The TR-3B Black Manta

The plasma, mercury based, is pressurized at 250,000 atmospheres at a temperature of 150 degrees Kelvin, and accelerated to 50,000 rpm to create a super-conductive plasma with the resulting gravity disruption.

The MFD generates a magnetic vortex field, which disrupts or neutralizes the effects of gravity on mass within proximity, by 89 percent... The current MFD in the TR-3B causes the effect of making the vehicle extremely light, and able to outperform and outmaneuver any craft yet.

...My sources say the performance is limited only by the stresses that the human pilots can endure. Which is a lot, really, considering along with the 89% reduction in mass, the G forces are also reduced by 89%. The crew of the TR-3B should be able to comfortably take up to 40Gs... Reduced by 89%, the occupants would feel about 4.2 Gs.

The TR-3Bs propulsion is provided by 3 multimode thrusters mounted at each bottom corner of the triangular platform. The TR-3 is a sub-Mach 9 vehicle until it reaches altitudes above l20,000 feet - then who knows how fast it can go!..."

I was skeptical of Mr. Fouche's claims when I first read them, as I'm sure that many of you are, but I was interested enough to do further research on what happens when you spin a plasma at high speeds in a ring (toroidal) configuration. I came across a physics article that described this exact configuration. The article said that, surprisingly, the charged particles of the plasma don't just spin uniformly around the ring, but they tend to take up a synchronized, tightly pitched, helical (screw thread) motion as they move around the ring.

The TR-3B Black Manta

This can be understood in a general way as follows: the charged particles moving around the ring act as a current that in turn sets up a magnetic field around the ring. It is a well-known fact that electrons (or ions) tend to move in a helical fashion around magnetic field lines.

Although it is a highly complex interaction, it only requires a small leap of faith to believe that the end result of these interactions between the moving charged particles (current) and associated magnetic fields results in the Helical motion described above. In other words, the charged particles end up moving in very much the same pattern as the current on a wire tightly wound around a toroidal core.

I thought that this was an interesting fact, but didn't see how it could possibly relate to antigravity, until I ran across the following article: "Guidelines to Antigravity" by Dr. Robert Forward, written in 1962 (available at: http://www.whidbey.com/forward/pdf/tp007.pdf).

Dr. Forward's article describes several little known aspects of Einstein's General Relativity Theory that indicate how moving matter can create unusual gravitational effects.

When I saw Figure 5 in Dr. Forward's article, the pieces of the puzzle all fell together. I instantly saw how the moving matter pattern that Dr. Forward describes as necessary to generate a gravitational dipole was exactly the same as the plasma ring pattern described in the physics article discussed above!

If Fouche's description is even close to correct, then the TR-3B utilizes this little known loophole in General Relativity Theory to create its antigravity effects!

The TR-3B Black Manta

Even though the TR-3B can only supposedly cancel 89% of gravity (and inertia) today, there is no reason why the technology can't be improved to exceed 100% and achieve true antigravity capability!

In theory, this same moving matter pattern could be mechanically reproduced by mounting a bunch of small gyroscopes all around the larger ring, with their axis on the larger ring, and then spinning both the gyroscopes and the ring at high speeds. However, as Dr. Forward points out any such mechanical system would probably fly apart before any significant antigravity effects could be generated.

However, as Dr. Forward states, "By using electromagnetic forces to contain rotating systems, it would be possible for the masses to reach relativistic velocities; thus a comparatively small amount of matter, if dense enough and moving fast enough, could produce usable gravitational effects."

The requirement for a dense material moving at relativistic speeds would explain the use of Mercury plasma (heavy ions). If the plasma really spins at 50,000 RPM and the Mercury ions are also moving in a tight pitched spiral, then the individual ions would be moving probably hundreds, perhaps thousands of times faster than the bulk plasma spin, in order to execute their "screw thread" motions. It is quite conceivable that the ions could be accelerated to relativistic speeds in this manner.

I am guessing that you would probably want to strip the free electrons from the plasma, making a positively charged plasma, since the free electrons would tend to counter rotate and reduce the efficiency of the antigravity device.

The TR-3B Black Manta

One of Einstein's postulates of GR says that gravitational mass and inertial mass are equivalent. This is consistent with Mr. Fouche's claim that inertial mass within the plasma ring is also reduced by 89%. This would also explain why the vehicle is triangular shaped.

Since it still requires conventional thrusters for propulsion, the thrusters would need to be located outside of the "mass reduction zone" or else the mass of the thruster's reaction material would also be reduced, making them terribly inefficient.

Since it requires a minimum of 3 legs to have a stable stool, it follows that they would need a minimum of 3 thrusters to have a stable aerospace platform.
Three thrusters, located outside of the plasma ring, plus appropriate structural support, would naturally lead to a triangular shape for the vehicle.

I was extremely skeptical of Mr. Fouche's claimed size for the TR-3B, of 600 feet across. At first, I thought that this must be a typo. Why would anyone in their right mind build a "Tactical Reconnaissance" vehicle 2 football fields long? They must be nuts!

However, the answer to this may also be found in Dr. Forward's paper. As Dr. Forward's puts it, "...even the most optimistic calculations indicate that very large devices will be required to create usable gravitational forces.

Antigravity...like all modern sciences, will require special projects involving large sums of money, men, and energy."

C) Third Description of this technology by Richard Boylan PHD

The TR-3B Black Manta

The **TR3-B 'Astra'** is a large triangular anti-gravity craft within the U.S. fleet. Black projects defense industry insider Edgar Rothschild Fouche wrote about the existence of the TR3-B in his book, Alien Rapture. The TR3-B does not depend solely or principally on its hydrogen-oxygen rockets. It is a highly reduced-gravity aerospace craft manufactured in secret "black programs" by Humans. The antigravity field produced reduces the vehicles weight by about 90% so that very little thrust is required to either keep it aloft or to propel it at Mach 9 speeds, or higher.

The TR-3B vehicle's outer coating is electro-chemical reactive and changes with electrical RF Radar stimulation and can change reflectiveness, radar absorptiveness, and color. This is also the first US vehicle to use quasi-crystals in the vehicle's skin. This polymer skin, when used in conjunction with the TR-3Bs Electronic Counter Measures and, ECCM, can make the vehicle look like a small aircraft, or a flying cylinder - or even trick radar receivers into falsely detecting a variety of aircraft, no aircraft, or several aircraft at various locations. A circular, plasma filled accelerator ring called the Magnetic Field Disrupter, surrounds the rotable crew compartment and is far ahead of any imaginable technology. Sandia and Livermore laboratories developed the reverse engineered MFD technology.

The plasma, mercury based, is pressurized at 250,000 atmospheres at a temperature of 150 degrees Kelvin, and accelerated to 50,000 rpm to create a super-conductive plasma with the resulting gravity disruption [reduction of almost all of the pull of gravity and effects of inertia]. The MFD generates a magnetic vortex field, which disrupts or neutralizes the effects of gravity on mass within proximity, by 89 percent. The MFD creates a disruption of the Earth's gravitational field upon the mass within the circular

The TR-3B Black Manta

accelerator. The mass of the circular accelerator and all mass within the accelerator, such as the crew capsule, avionics, MFD systems, fuels, crew environmental systems, and the nuclear reactor, are reduced by 89%.

The current MFD in the TR-3B causes the effect of making the vehicle extremely light, and able to outperform and outmaneuver any craft yet constructed - except, of course, those back-engineered total-antigravity craft which the government does not admit exist.

The TR-3B is a high altitude, stealth, reconnaissance platform with an indefinite loiter time. Once you get it up there at speed, it doesn't take much propulsion to maintain altitude. With the vehicle mass reduced by 89% the craft can travel at Mach 9, vertically or horizontally. My sources say the performance is limited only by the stresses that the human pilots can endure. Which is a lot, really, considering along with the 89% reduction in mass, the G forces are also reduced by 89%. The crew of the TR-3B should be able to comfortably take up to 40Gs.

The TR-3Bs propulsion is provided by 3 multimode thrusters mounted at each bottom corner of the triangular platform. The TR-3 is a sub-Mach 9 vehicle until it reaches altitudes above 120,000 feet - then who knows how fast it can go!

The reactor heats the liquid hydrogen and injects liquid oxygen in the supersonic nozzle, so that the hydrogen burns concurrently in the liquid oxygen afterburner. The multimode propulsion system can operate in the atmosphere, with thrust provided by the nuclear reactor, in the upper atmosphere, with hydrogen propulsion, and in orbit, with the combined hydrogen/oxygen propulsion. The engines are reportedly built by Rockwell.

The TR-3B Black Manta

Diagram from Lockheed Martin's Compact Fusion Patent

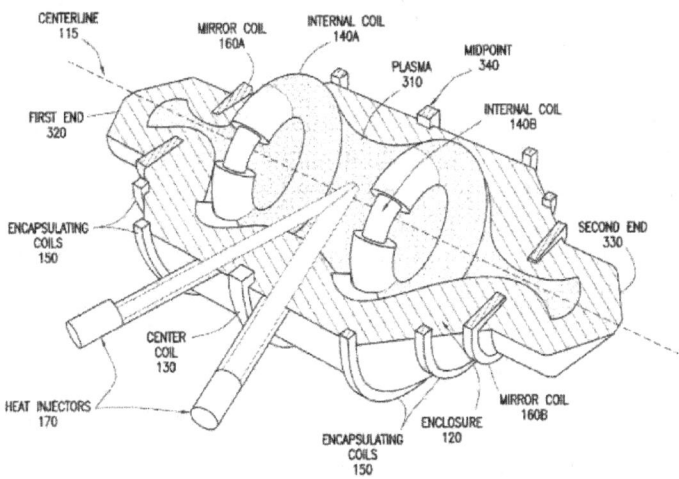

But that's only the beginning. The most extreme (and perhaps, unbelievable) aspects of the TR-3B could potentially be explained by the work of one aerospace engineer who is reportedly now employed by the U.S. Space Force.

Does Salvatore Pais have the keys to a UFO?

In 2020, the U.S. Navy filed a series of extremely unusual new patents pertaining to the sort of seemingly science-fiction technology found in claims about the TR-3B. These patents cover a wide range of topics from gravitational-wave generators to space time modification weapons, and they all have one thing in common. Or rather… one *person*. Dr. Salvatore Cezar Pais.

According to Pais' official biography, he's spent time as a NASA graduate research fellow, an advanced concepts

The TR-3B Black Manta

analyst with Northrop Grumman (the same firm reportedly behind the TR-3A), the Department of the Navy Strategic Systems Program, and now, according to reports, he's with the U.S. Space Force.

So far, it's been confirmed that the Navy has poured at least $466,000 into helping advance Pais' work since 2017. Much of the work has subsequently been made public in the form of three patents filed by the Navy, the first of which came in 2016 with the title: *Craft using an inertial mass reduction device*.

Craft using an inertial mass reduction device
US · US10144532B2 · **Salvatore Cezar Pais** · The United States Of America As Represented By The Secretary Of The Navy
Priority 2016-04-28 · Filed 2016-04-28 · Granted 2018-12-04 · Published 2018-12-04
A craft using an inertial mass reduction device comprises of an inner resonant cavity wall, an outer resonant cavity, and microwave emitters. The electrically charged outer resonant cavity wall and the electrically insulated inner resonant cavity wall form a resonant cavity. The microwave emitters ...

Piezoelectricity-induced High Temperature Superconductor
US · US20190348597A1 · **Salvatore Cezar Pais** · United States Of America As Represented By The Secretary Of The Navy
Priority 2017-08-16 · Filed 2019-07-23 · Published 2019-11-14
The present invention is a high temperature superconductor comprising of a wire, which comprises of an insulator core and a metal coating. The metal coating is disposed around the insulator core, and the metal is coating deposited on the core. When a pulsed current is passed through the wire, ...

High frequency gravitational wave generator
US · US10322282B2 · **Salvatore Cezar Pais** · The United States Of America As Represented By The Secretary Of The Navy
Priority 2017-02-14 · Filed 2017-02-14 · Granted 2019-06-18 · Published 2019-06-18
A high frequency gravitational wave generator including a gas filled shell with an outer shell surface, microwave emitters, sound generators, and acoustic vibration resonant gas-filled cavities. The outer shell surface is electrically charged and vibrated by the microwave emitters to generate a ...

If that wasn't shocking enough, here's how a Navy PowerPoint slide labeled "For Internal Use Only" explains the implications of Dr. Pais' technology, as revealed by a Freedom of Information Act request filed by The Warzone.

The TR-3B Black Manta

> ### Potential Uses
> - What is the Navy's potential use for this invention:
>
>   - Imagine our Navy's ships, submarines aircraft and (Marine Corps) armored ground vehicles being powered with safe, reliable, virtually limitless fusion energy. Imagine the power of the Sun confined in a compact, relatively small space. With the Plasma Compression Fusion Device (PCFD), this figment of imagination becomes a tangible reality.
>   - The present invention can produce power in the Gigawatt to Terawatt range (and higher) with input power in the Kilowatt to Megawatt range, and possibly lead to Ignition plasma burn.
>   - Under uniquely defined conditions, the Plasma Compression Fusion Device can lead to development of a <u>Spacetime Modification Weapon</u> (**SMW**- a weapon that can make the Hydrogen bomb seem more like a firecracker, in comparison). Extremely high energy levels can be achieved with this invention, under pulsed ultrahigh current (I) / ultrahigh magnetic flux density (B) conditions (Z-pinch with a Fusion twist).
>   - **SMW Energy Yield ~ $I^2 B^3$**
> - **Is there the potential for commercial use – YES**
>   - The design of Thermonuclear Fusion Reactors (safe, reliable, limitless energy) for commercial electricity generation.
>   - The design of Fusion-driven Aircraft Jet Engines.
>   - The design of Fusion-induced Intergalactic Space Drives.
>
> FOUO- PAX 285 - Plasma Compression Fusion Device ▬▬▬   5.

U.S. Navy slide mentioning the Space-time Modification Weapon, released via FOIA filed by The Warzone.

If you're starting to get awfully skeptical of the things I'm telling you, I don't blame you. I'm skeptical too – but Pais does have some very noteworthy supporters, including the Chief Technology Officer of the U.S. Naval Aviation Enterprise, Dr. James Sheehy.

"China is already investing significantly in this area," Sheely told Patent Examiner Philip Bonzell, and "would prefer we [the U.S.] hold the patent as opposed to paying forever more to use this revolutionary technology" as he asserts "this will become a reality." *Docs Show Navy Got 'UFO' Patent Granted By Warning Of Similar Chinese Tech Advances from* The Warzone

The TR-3B Black Manta

The TR-3B Black Manta

9.0 Aircraft Sightings

British sightings and UK Ministry of Defense Report

"An example UAP formation of the triangular type," depicted in a Technical Memorandum on the subject of UAP commissioned by the British government.

A declassified report from the UK Ministry of Defense, addressing Unidentified Aerial Phenomena (UAP) within the UK Air Defense Region and code named Project Condign, includes analyses of black triangle sightings.

The report includes the statement that "the majority, if not all, of the hitherto unexplained reports may well be due to atmospheric gaseous electrically charged buoyant plasmas" that are "capable of being transported at enormous speeds under the influence and balance of electrical charges in the atmosphere." The report also notes that "at least some" of the black triangle observations likely arise from meteor entry into the atmosphere.

The TR-3B Black Manta

Regarding the triangular shapes, the report also states: "Occasionally ... it seems that a field with, as yet, undetermined characteristics, can exist between certain charged buoyant objects in loose formation, such that, depending on the viewing aspect, the intervening space between them forms an area (viewed as a shape, often triangular) from which the reflection of light does not occur. This is a key finding in the attribution of what have frequently been reported as black 'craft,' often triangular and even up to hundreds of feet in length."

A recommendation in the report is that no attempt be made on the part of aircraft to intercept or outmaneuver these objects, and instead to place them astern to mitigate the risk of collision. The report also speculates that the hypothesized plasma formations, through their "magnetic, electric or electromagnetic" fields, could have the potential to induce in observers vivid, but mainly incorrect, perceptions.

The Project Condign report was not peer-reviewed, and some authors doubt its scientific veracity.

Individual sightings

1980s Hudson Valley sightings

During the early 1980s, several hundred people claimed to have witnessed UFOs flying over, or near to, the Hudson River in New York State. These sightings involved hovering or slowly flying V-shaped objects rimmed with colorful lights. Several pilots claimed responsibility for these UFOs, reporting that the objects, some tracked to a local airport and parking lot, were ultralight aircraft flown in formation.

The TR-3B Black Manta

1989–1992 Belgian wave

The Belgian UFO wave began in November 1989. The events of 29 November were documented by over thirty different groups of witnesses and three separate groups of police officers. All of the reports related to a large object flying at a low altitude. The craft was flat and triangular, with lights underneath. This giant craft did not make a sound as it slowly moved across the landscape of Belgium.

The Belgian UFO Wave of 1989–1992 – A Neglected Hypothesis discusses some sightings that helicopters can explain. Most witnesses reported that the objects were silent. This report argues that the lack of noise could be due to the engine noise in the witnesses' automobiles or the strong natural wind blowing away from the witnesses.

Black triangle UFOs have been claimed to be visible to radar. During the 1989–1990 Belgian UFO wave, two Belgian Air Force F-16s attempted to intercept an object detected by radar, but the pilots did not report seeing an object. This entire Belgian UFO wave, however, has been disputed by skeptics.

Phoenix Black Triangles

A widely reported appearance(s) of black triangles involved the "Phoenix Lights" events, during which multiple unidentified objects were observed near Phoenix, Arizona and videotaped by both the local media and residents beginning on Thursday, March 13, 1997. Some observed objects/lights appeared to be grouped in a large "V" formation that lingered for several minutes. Some residents reported one of the black triangles to be over a mile wide and that it drifted slowly over their houses, blocking out the night sky's stars.

The TR-3B Black Manta

An official report from the US Air Force concluded that the military had been locally testing aircraft-launched flares during that period.

2000 Southern Illinois incident

The "St. Clair Triangle", "UFO Over Illinois", "Southern Illinois UFO", or "Highland, Illinois UFO" sighting occurred on January 5, 2000 over the towns of Highland, Dupo, Lebanon, Shiloh, Summerfield, Millstadt, and O'Fallon, Illinois, beginning shortly after 4:00 am. The incident was featured in several television shows including Seeing is Believing, a Discovery Channel special UFOs Over Illinois, and an episode of the Syfy series Proof Positive. Sufjan Stevens included this incident in the song "Concerning the UFO Sighting Near Highland, Illinois" from his 2005 album Illinois. The FAA said sighting reports may have been due to an advertising blimp operated in the area by the American Blimp Company.

2004–2006 Tinley Park Lights

Three red lights hovered in a triangular formation were seen by multiple witnesses in Tinley Park and Oak Forest, Illinois, on August 21, 2004, two months later on October 31, 2004, again on October 1, 2005, and once again on October 31, 2006. Some witnesses photographed the lights and captured them on video. According to some urologists, the video evidence suggests that the lights kept the geometrical shape and moved as if they were attached through a dark object. The incident was examined in a Dateline NBC episode on May 18, 2008, and in the episode "Invasion Illinois" of the television series UFO Hunters premiered on The History Channel on October 29, 2008.

The TR-3B Black Manta

2008 Stephenville, Texas

Around 8 January 2008, there was a mass sighting in Stephenville, Texas. In the 2023 Netflix documentary series Encounters, it is claimed that there were black triangles and inside were what looked like insects or a praying mantis.

Military aircraft

Classified military aircraft may be responsible for many black triangle UFO reports. Several such sightings have been reported over Antelope Valley, an area of desert in southern California. This stretch of desert draws people interested in potential "black project" aircraft because it is close to several known military research and testing areas, such as Edwards Air Force Base in California and United States Air Force Plant 42. A geographic analysis by the now-inactive National Institute for Discovery Science suggested that black triangles might be U.S. Air Force craft.

At least some of the proposed military types may be fictitious. The Northrop TR-3A Black Manta is a speculative surveillance aircraft purported to belong to the United States Air Force and to have been developed under a black project. It was said to be a subsonic stealth spy plane with a flying wing design. It was alleged to have been used in the Gulf War to provide laser designation for Lockheed F-117 Nighthawk bombers, for targeting to use with laser-guided bombs (since the F-117 possesses a laser designator, the reason for both aircraft being utilized is unclear). There is little evidence to support the TR-3's existence; however, it is possible that black triangle UFO reports associated with Black Manta could be a technology demonstrator for a potential new-generation tactical

The TR-3B Black Manta

reconnaissance aircraft, and/or that TR-3 refers to a Technical Refresh of an existing program.

Geoscientist Ben McGee has identified border patrol drones with infrared anti-collision or identification lights to explain some black triangles.

There are hundreds more sightings around the world. A map of the United States showing sightings is in chapter three.

The TR-3B Black Manta

10.0 Actual Photographs

There are hundreds of triangular black manta craft photos. These photos all show a triangular aircraft with lights (or propulsion) on each corner and one in the middle.

This craft can also achieve tremendous and impossible speeds. This is probably because this ship is supposed to reduce gravity by over ninety percent so any beings inside of it will only experience one tenth of gravitational forces from violent maneuvers.

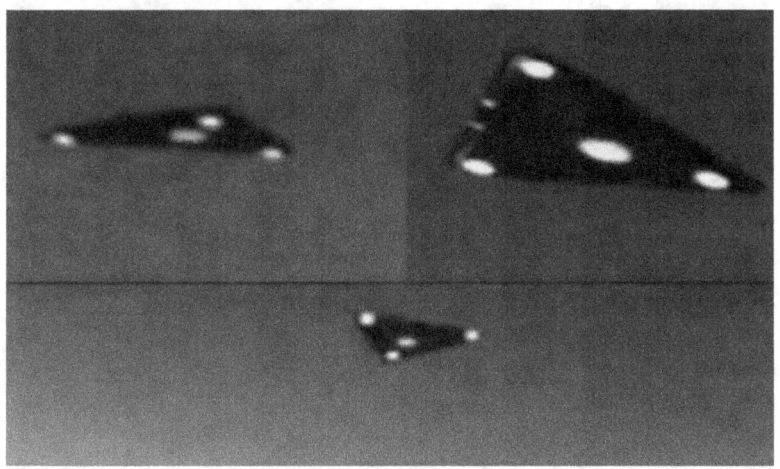

The TR-3B Black Manta

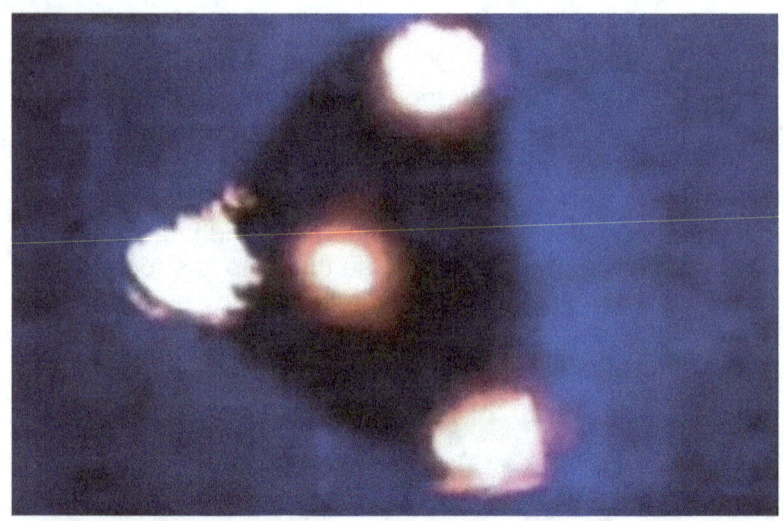

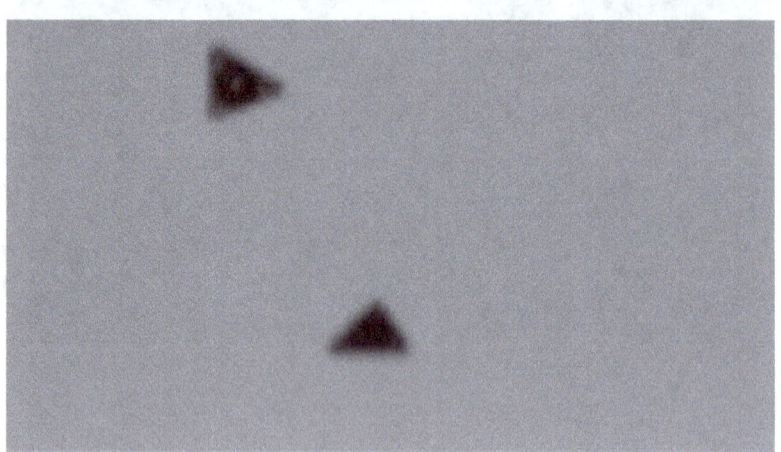

The TR-3B Black Manta

Paris, France 2009

The TR-3B Black Manta

In the presumed picture, taken by a classified STS mission in 2002, you can see a full-body shape aircraft (a black triangle) coming out from Earth's atmosphere to go into space. Could it be the Aurora or the TR3-B Astra? Or an alien craft? Or fake?
Courtesy: C. Barbato.

The TR-3B Black Manta

11.0   Claims of this Aircraft in Desert Storm

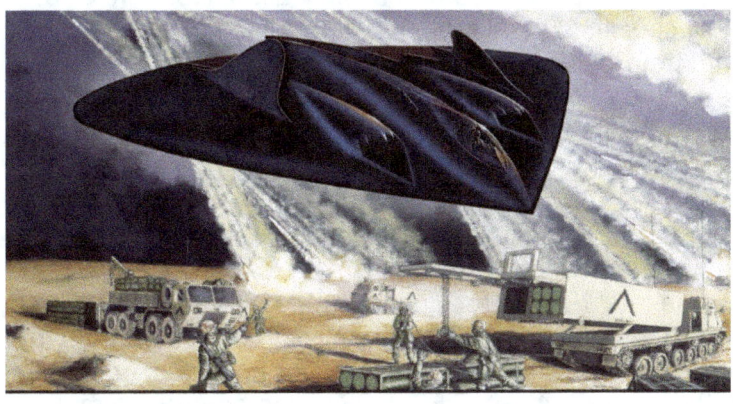

For decades, rumors have swirled about the United States secretly operating highly classified black, triangular craft known as the TR-3A and TR-3B, with some going as far as to claim that these platforms were designed using reverse-engineered alien technology. In fact, in the early '90s, it was even reported that these covert craft flew alongside the F-117 during combat operations over Iraq in *Desert Storm*.

<u>Could these Black Triangles have been secret American aircraft?</u>

The US has placed a heavy emphasis on aviation technology since the very inception of manned flight, with the U.S. Army placing an order for the world's first military aircraft from the Wright Brothers in 1908. Today, America's warfare doctrine leans heavily on the nation's ability to take and *keep* control of the airspace over any battlefield the world over. Of course, maintaining that capability in the face of increasingly capable international competitors has always required both significant *investment* and equally significant *secrecy*.

The TR-3B Black Manta

You can find a laundry list of secret aircraft programs that, once disclosed, still seemed *awfully alien*. Not only were highly classified stealth aircraft like the F-117 flying for years before the government acknowledged it existed, but even *more exotic* secret aircraft are now known to have been prowling the skies over the Southwestern United States for years.

Although defense spending did see consistent reductions following the fall of the Soviet Union, it's worth noting that, until the late 1990s, the United States was still allocating a larger percentage of the nation's GDP to defense than it does today. In fact, when adjusted for inflation, America's 1992 defense budget of $325.03 billion equates to more than $718 billion today – meaning Uncle Sam certainly had the money to fund a variety of classified programs. Further, in 1991 it was reported that the U.S. Air Force had devoted more than $60.3 billion to classified research, development, and procurement over the five preceding years – that's the equivalent of nearly $137 billion today, or enough to purchase more than 1,500 F-35As in today's market.

Reports of America's TR-3A Black Triangle Serving in Desert Storm

In 1991, America's Black Triangle was seemingly revealed to the world in a series of articles published by Aviation Week and Popular Mechanics. According to Aviation Week, the stealthy aircraft was designed by Northrop – the same firm responsible for the black, triangular B-2 Spirit – in 1976 alongside Lockheed's Have Blue efforts that would ultimately produce the F-117. Northrop called its stealthy triangular aircraft the Tactical High Altitude Penetrator (THAP).

The TR-3B Black Manta

Pictures from a YouTube video of a TR-3B over a town in Iraq and then the explosions over the town it created:

According to Aviation Week, the Air Force ultimately awarded Northrop a fixed-price research and development and demonstration/validation contract near the end of 1978 to build a prototype high-altitude reconnaissance aircraft based on their THAP design. That prototype, Aviation Week claimed, made its first test flight out of Area 51 in 1981, and a production contract was subsequently awarded in 1982.

The TR-3B Black Manta

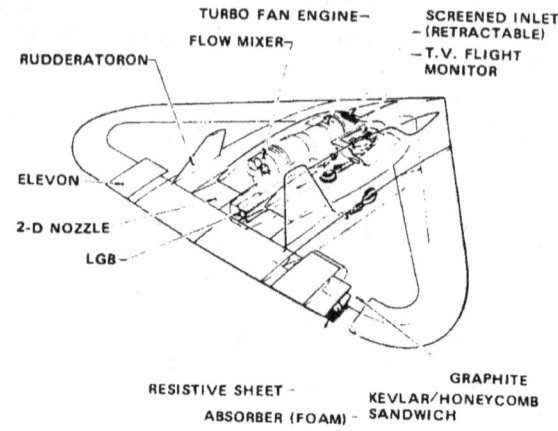

Shown in the photograph above is the Tactical High Altitude Penetrator (THAP) aircraft whose design is optimized to avoid detection by hostile defenses.

(SecretProjects.uk)

In a follow-up article, Aviation Week went on to claim that Northrop's TR-3A was about 42 feet long, 14 feet high, and had a wingspan of 60-65 feet, which describes a much smaller aircraft than popular reports made the "TR-3B" to be. Yet, this would seem to be in keeping with a sighting that is often attributed to Aurora over the North Sea in 1989, reported by a trained airfield observer named Chris Gibson.

The TR-3B Black Manta

Manipulated image meant to represent Chris Gibson's sighting. (Twitter)

According to Aviation Week's unnamed sources, these aircraft "may have" been deployed to Alaska, Britain, Panama, and Okinawa, as well as flying in concert with the F-117 Nighthawk during combat operations in Iraq to provide laser-designation of targets over Baghdad.

That claim, while not officially substantiated, might explain why the documents given to Iraqi MiG pilots to identify the F-117 in the air also showed the silhouette of the B-2.

Confusion over *just what was being seen* in the skies over Iraq may have prompted them to include the only other black triangle aircraft America was known to fly. However, the B-2 was not in service then, which would raise the question of what they actually saw. Though, admittedly, this line of reasoning may be a bit of a stretch.

The TR-3B Black Manta

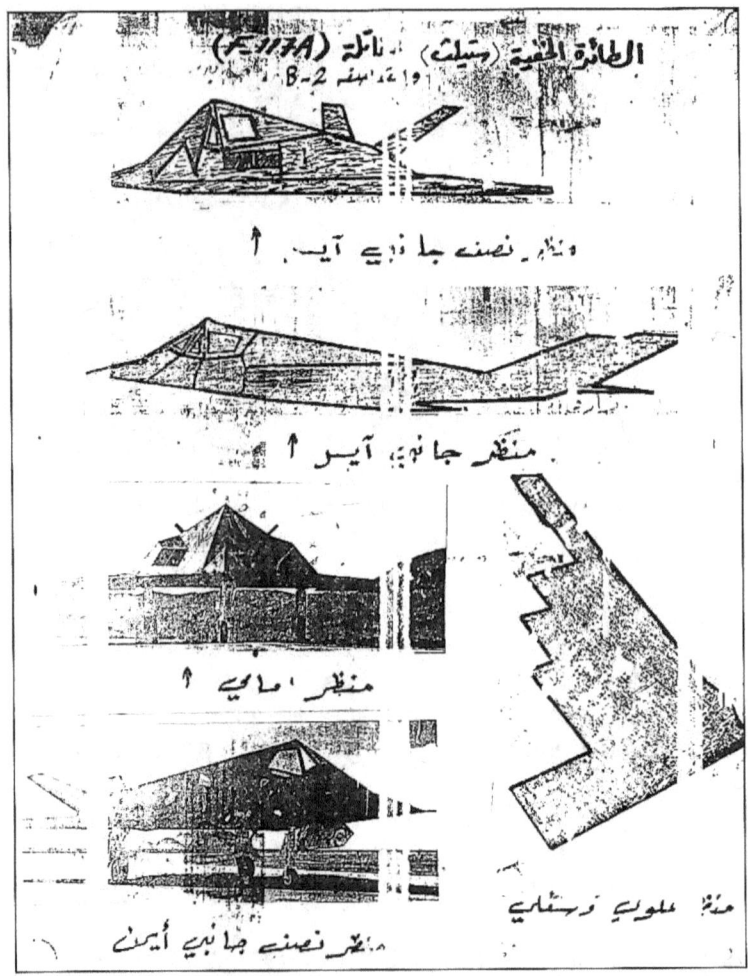

Document recovered from Iraqi Air Force Base in Tallil.

Popular Mechanics discussed the Northrop TR-3A alongside other undisclosed but *reportedly sighted* aircraft in their coverage, including another *boomerang-shaped* platform said to be completely silent and boasting a massive wingspan that stretched between 600 and 800

The TR-3B Black Manta

feet – or three to four times the size of the B-52 Stratofortress.

Artwork of the TR-3A in Iraq featured in Popular Mechanics in 1991

Like Aviation Week, Popular Mechanics also reported on the TR-3A being significantly quieter than other aircraft, but not silent, as is often reported about the TR-3B.

But evidence of TR-3A's existence isn't quite as strong as it seems

Those articles in Aviation Week that so authoritatively relayed the story of the TR-3A's development? Well, they were both written by or with the support of William Scott – a journalist who is now known for sometimes getting a little too excited about the unusual topics he covered, resulting in some serious, if likely unintentional, stretching of the truth.

The TR-3B Black Manta

In 1990, for instance, he reported that the United States had a secret hypersonic bomber that could launch nuclear weapons from vertical launch tubes. That aircraft, of course, never manifested either.

The TR-3 designation, many now believe, was the result of Scott simply *mishearing* stories about *Tier 3*, which was a program that followed Tier 2 (an effort that resulted in the Global Hawk drone). Tier 3 was supposed to be an unmanned SR-71 successor that was also known as "Quartz," but that ultimately didn't make it beyond the design stage. Elements of the Tier 3 program, known as Tier 3 Minus, did ultimately result in Lockheed and Boeing building the Darkstar in 1996 – no, not the hypersonic one Maverick flew, but rather a much slower drone meant for ISR duties.

In fact, when you save this image of the Darkstar from Wikimedia Commons, the file name includes *both* Darkstar and "Tier 3."

And when you read the Popular Mechanics coverage that was published in 1991, you'll find that it pulls *primarily* from Scott's reporting in Aviation Week.

Many, including *me*, were struck by the details in the Aviation Week story because it's a well-respected outlet with a history of having insider information. But the outlet also has a well-recorded history of publishing some less-than-factual accounts of black aircraft programs over the years; stories with little in the way of disclosed sources that made lofty claims about a near-term future that never manifests.

The TR-3B Black Manta

In 2006, space historian and policy analyst Dwayne Day summed up how academics now perceive Scott's 1990 coverage of the TR-3A:

"The Manta story demonstrates a pattern that Scott repeats in all of his black airplane stories. Usually there is a small bit of real information about a classified aircraft project. Scott then connects alleged sightings of an unusual aircraft in flight to this bit of information. Then the article is padded out with a large amount of speculation, usually involving various studies and research projects conducted by various contractors. The characteristics are always the same, however: he never quotes anybody by name who has any direct connection to the alleged program, and he never even includes anonymous quotes of anybody who supposedly knows the big picture about the alleged program."

The TR-3B Black Manta

The TR-3B Black Manta

12.0   The Silent Triangle

Many people who have seen the TR-3B claim it is totally silent which would be consistent with it not using conventional engines but some type of anti-gravity technology and only small thrusters for movement.

<u>Sighting by Alex Hollings (Author and news writer)</u>

Sometime between 1994 and 1995, I was living on County Road in Torrington, Connecticut. It was a summer evening, and my older brother was having a friend sleep over that night. In an unusual bit of brotherly generosity, the older kids were letting me hang out in the family room with them as we watched TV and ate pizza my parents ordered – in a real way, it was about as good a night as 10-year-old Alex had ever had.

Sometime in the evening, my brother asked me to run out to my father's old Lincoln Town Car to get the pack of playing cards we kept in the glove box for road trips. I was so happy to be included in the night's fun that I didn't think twice, and took off out the side door, down two flights of concrete steps, and into our driveway. But as I opened the passenger side door on my dad's big green land-yacht, the hair on the back of my neck stood up, as though someone was standing right behind me.

I spun around to see who was there… but found nothing but empty space between me and the neighbor's house across the street. And then suddenly a massive and utterly silent black triangle made its way over the treeline behind the house across the street. It was dark colored – but a lighter shade of grey than the night sky it hid behind it, blocking the starlight as it passed. I couldn't tell you what its altitude was, but I knew it was flying low – too low to be

The TR-3B Black Manta

any normal aircraft. But the thing that left me most unnerved was its lack of audible engine noise.

As the triangle passed over me, I stood frozen for a second before the panic set in and sent me running (and screaming) back inside. My brother and his friend met me at the door and before long, my parents were joining them in reassuring me that all I saw was a low-flying airplane.

compGeniusSuperSpy (From Reddit)

i saw a black triangle ufo in 1994 in sanger, CA. it was the size of a stadium, utterly silent, and it was only at about 1500 ft altitude. it glided slowly, in the middle of the night. it was not a tr-3b and there were no lights to speak of. it was opaque and matte.

Honestly it framed the entire way i see the world. Luckily I was a child with a malleable brain or it might have caused me to lose my mind.

Absolutely without a single doubt was not of this earth.
I remember it like it was yesterday.

averagemaleuser86 (From Reddit)

Mother fucker. Now I know what I saw the night of Y2K in Warner Robins, GA going directly towards WRAFB!... we thought it might have been a B2 or a Nighthawk. We were kids... waiting on the chaos of y2k that never happened. But just happened to look up and there it was just gliding slowly and straight. Big black triangle shape. Very small lights on the tips. No sound at all... could hear the traffic at the intersection about a mile nearby. It was close to midnight... again we were outside because we thought chaos was gonna happen and the lights were gonna go

The TR-3B Black Manta

out and such. This is wild now that I think I can explain it!! The other thing is, I now work out here at RAFB, and there's nothing secretive out here. So really don't understand what it would be doing heading here. But that was 23 years ago! Never know.

Anonymous (From Reddit)

I saw something very similar in 2000 or 2001. Was living in Fort Leonard Wood, MO at the time. I woke up in the night, went to the bathroom, and then came back to bed. As I dropped into my bed, I glanced out the window and saw a massive black triangle. Similar light layout to that pic. It just hovered there over the forest and was completely silent. I starred at it for what felt like forever but I'm guessing was only a couple dozen seconds. Then it flew off in an instant. No acceleration. Just an instant blur forward and it was gone.

LiquorThenLickHer (From Reddit)

Saw the same thing in 2009 outside of DC with a buddy. Night time, stadium sized, silent, hovering, low to the ground. But it did have a light at each point. We jumped in the car and tried following it when it started moving. Kept up for about a mile or so taking different roads that didn't block the view. My buddy saw it leave, I was driving. Said it like zig zagged or something then was gone extremely fast. Me and my buddy still talk about it till this day, will never forget it.

The TR-3B Black Manta

The TR-3B Black Manta

13.0 Summary

So what conclusions can we draw from the stories and evidence presented in this book?

- This craft goes by various names including TR-3B, Black Manta, and Astra
- That these triangular craft do exist as numerous pictures attest.
- Several sources say that the propulsion technology is based on anti-gravity from a rotating toroidal high pressure mercury vapor which acts as a "Gravity Shield" to cut effective gravity by 90 percent.
- The origins of this craft can be traced back to the 1940s and 1950s from captured alien technology
- Many say the size is over 600 feet in diameter and it is totally silent. Observers report there was no sound when they made their sightings.
- Several pictures show that this Astra craft can also go into space.
- It may have been used to destroy enemy objectives during Operation Desert Storm.

Hope you enjoyed this book and learned something new about the TR-3B.

All the Best,

Martin K. Ettington
December 2024

The TR-3B Black Manta

14.0   Bibliography

Ettington, M. K. (2021). *Human & Alien UFO Anti-Gravity Research.*

Hollings, A. (2023). *https://www.sandboxx.us/news/airpower/exploring-the-claims-that-americas-tr-3a-ufo-fought-in-desert-storm/.* Retrieved from https://www.sandboxx.us.

https://archive.org/details/TopSecretTr-3bAstraDocuments/100ProofTheMilitaryHasAlienTr-3bCraftToTravelIntoSpace/. (2022). Retrieved from Astra Documents.

https://medium.com/predict/eugene-podkletnovs-new-gravity-modification-experimental-video-b7813b04c6f8. (2022). Retrieved from Eugene Podletnov's New Gravity Modification Experiments.

https://www.youtube.com/watch?v=wkm3KxIyZDE. (2024). Retrieved from Black Triangle UFOS and links.

blueberry. Mushrooms are 90% water weight, which is why they are even used in oil spills for their absorption properties. If you don't plan on eating your mushrooms fresh, then begin the drying process immediately and don't waste too much time letting them just sit around, or they will begin to rot.
When drying, you can rest them on paper towels and run a fan on low in front of them. You will know when they are completely dry because they will be brittle like a cracker. You can also eat them fresh if you choose, they are supposedly way stronger this way. If you get them completely dry, then you can store them in ziplock bags for a long time. Once they are completely sealed, mushrooms can be stored in a cool, dark area for several months.

A good way to determine how much your dried mushrooms will weigh, is by weighing them immediately after picking, and your actual weight will be ten percent of that number. When your mushrooms are completely dry, they will be very brittle, similar to a cracker. That is what you want, don't mistakenly think they are dry, bag them and then they turn bad.

After you have completed these steps you can continue to rinse and repeat exactly for at least three or four more flushes. When you become experienced with this you can try new techniques, such as grain to grain transfer, mixing two types of spores into one substrate (essentially creating a hybrid strain) and making your own spore syringes and spore prints.

Now these tricks are admittedly not my specialty, I typically order new syringes when it's time to start over. That being said I know the benefits to these techniques. Grain to grain transfer can be completed by taking a little bit of a fully colonized cake and introducing the colonized grain into a fresh batch of grain. This can be done infinitely, which means no more purchasing spores. So very similar to the injection of spores, you are just adding the mycelium by hand.

Similar to injecting the spores, the colonized grain will begin to slowly start colonizing the new grain, at an even faster rate due to the head-start. This is a very touchy method and can be easy to mess up attempting, so be very clean and watch multiple videos in order to get it down. Studying is key. If you find yourself mastering the indoor process you could also adventure outdoors, as there are many ways to accomplish this.

Another way to save a great amount of money on future projects would be spore prints, which you can also make yourself. After the cap opens in the late stages of growing, you can use a small piece of aluminum foil to collect the spores as they drop, using a small, clean metal scraper to gently scrape the spores from inside the cap. After this step is complete you can add the spores to a syringe with distilled water and now you have your own spore syringe.

When I first began growing (about ten years ago) I started with jars, and stuck with jars for the first two years. I enjoyed this method for a while, the yields were not very large (maybe 15 grams per flush, using 12 jars) but the process was quite a bit faster than any alternatives. Shortly into my second year I began experimenting with bags, which I found rather enjoyable after the yield was complete.

Each bag that went through successfully produced about 15 grams per flush on average, which is equivalent to all 12 jars. Once you master these steps you can easily produce several pounds every few months in your closet. If you time your cycle correctly, you can be producing yields monthly by simply growing jars and bags together.

I found the best success for indoor growing is using both, spawns bags and substrate jars. Using both methods will ensure you have speedy results with the jars, and a heavy yield with the bags. This will also give you experience with both so you can determine what's best for you, or you may just prefer using both. If you become good with the bags, you can experiment with much larger bags, which means a much larger yield.

If you're interested in bulk growing then check out bulk growing techniques on YouTube for free. Remember you can study commercial growing for tips from bulk growing professionals. This guide is best for personal use, hence it's a beginners guide that takes place in a closet grow room. That being said the techniques learned here can be used on a larger scale with ease.

When studying the commercial growers don't be afraid to experiment with anything that seems feasible, these commercial mushrooms are grown exactly the same way as psilocybin mushrooms, just on a grand scale. You can check out plenty of pictures readily available for free online which documents commercial grow rooms.

The commercial grow rooms are very similar to the room that was just built using this book, just in a very advanced high tech way. They use the same sanitizing U.V lights and they typically hang the substrate bags upside down, or on shelves. The process is very neat to check out and can give you plenty of ideas for your own room, so don't be afraid to experiment.

The larger the projects you take on, the more you risk contamination. Be safe while growing and never look the other way when it comes to any types of mold, if it's not pure white mycelium growth toss it. Keep in mind mushrooms will grow on a contaminated cake, and they will also make you VERY sick, or whoever you give them too. If you make a mistake, no big deal, try again. Don't risk getting sick or worse on a contaminated cake.

If you decide to adventure growing outdoors, then you won't have to worry as much about contamination, but always use a mushroom identifying book if you pick wild mushrooms. Now I have covered that the drug present (Psilocybin) in magic mushrooms will bruise (turn blue) when damaged or exposed to air, but that doesn't mean that a wild mushroom doesn't contain psilocybin and poison.

Basically if you aren't a mushroom identifying expert, then stick to growing at home, with spores that you've ordered from a reputable company. If you have the resources, then you can use horse manure, straw and mulch and inoculate them outside and let nature run its course, some people report great success this way.

Another great way to get the most bang for your buck is with your spent (used-up cakes) after they have been flushed multiple times and throw them in your garden, you may just be surprised with what pops up. I have done this several times and always had at least a dozen or so mushrooms appear within a week or two. If you have any mulch in a shady spot of your yard, then this will be great for planting your cakes.

This is a never-ending experiment, you can constantly alter these steps and have different results because of it. Every grower does something a little different and unique, you should too.

Find what works best for you, maybe you will discover something new, such as developing a new strain or mastering making spore syringes or prints.

You can also try your luck with liquid culture, a tricky technique that is usually only attempted by seasoned growers. Liquid culture consists of a sterilized nutritious solution, usually a mixture of water and various kinds of sugars, which has been inoculated with fungus spores or mycelium. Once colonized, liquid culture is used to inoculate PF jars or Grain Spawn.

The advantages of using liquid culture instead of spores to inoculate your jars or spawn are that the cultivator no longer has to wait for spores to germinate, thus reducing incubation times by at least a week, and that 1cc or spore solution can be used to create an effectively limitless amount of liquid culture.

The options are unlimited in many ways and the entire process is very rewarding, even more so than growing marijuana in my opinion because less people have grown mushrooms, making the experience more unique. Also, it is very strange to see the way these things will grow, two and even three headed mushrooms and sometimes hundreds of tiny ones will grow, or one giant one.

If you really want to experiment, you can try your luck growing rate and exotic strains. This can be very rewarding, but much more difficult and I wouldn't recommend doing this immediately. However if you do want to grow something more on the exotic side (Z strain, Albino, UFO etc) be sure to check the optimal conditions for that specific strain, as they will vary.

Mushrooms are very unpredictable in their growing stages, and in their potency. You can pick two mushrooms right next to each other and one can be ten times as strong as the other. Another factor that can determine potency is how early you pick them (earlier being more potent), with the strongest being the aborts (mushrooms that began to grow then aborted, resulting in very tiny mushrooms with dark blue caps).

Mushroom aborts are usually around ten times more potent per gram, so be very careful if you ingest abort in a large quantity. If you're also new to ingesting mushrooms, I suggest starting with under two grams. If you begin with more than two grams for your first experience, it may turn out bad. It's always best to take too little, rather than take too much.

If you want to collect the spores, then you will have to let the caps open, which will result in slightly less potent mushrooms, but more weight. This is because the psilocybin will spread throughout the mushroom, as the mushroom grows larger it doesn't produce more psilocybin. The caps contain roughly twice the psilocybin content levels as the stems do, which is why most people ask for the caps while eating mushrooms.

If you're like the average person, you likely won't enjoy the taste of psilocybin mushrooms, they aren't very pleasant. I usually grind my mushrooms into a powder, then mix the powder into a tea, creating mushroom tea. You can also put the powder into empty pill capsules and ingest them that way, or with chocolate or peanut butter.

You can also experiment with microdosing, which has become quite popular recently. Microdosing consists of ingesting small amounts of the drug, resulting in heightened senses, without the hallucinations. Some athletes even report doing this before major games, or in some cases before boxing and MMA matches.

Hopefully you now have the confidence to attempt tackling this project head-on, and if you become good enough you can have a career in mycology. There are plenty of career opportunities in this field for those who take it seriously, especially if you can make mass amounts of spore prints or live culture.

If you can accomplish this in a clean environment, then there are opportunities to sell the live culture and or spore prints to the reputable companies that you bought them from initially when you first began. This is because there are times in the season that the demand is greater than the supply, resulting in these companies looking elsewhere for the product.

This may seem far-fetched, but it's very possible after you have the correct amount of experience and transactions with these companies. Most of the companies I've encountered are more family owned businesses, which makes them more accessible for selling to, even most of their web pages acknowledge that they purchase live culture and or spore prints.

I truly hope you've enjoyed this guide, and I also wish you luck in your mushroom journey. This information is everything I've learned about mushrooms in the last decade, from endless research on the topic, along with countless grows of my own. If you read the book in its entirety, and stuck to the basics, then you should be on your way to growing professional grade mushrooms in no time.

Never get discouraged in your grows, hope for the best and expect the worst. Look at the entire process as a learning experience, and you either have a good yield or learn something for the next attempt. You'll be amazed how quickly you pick this hobby up, and how potent of a psilocybin mushroom you can grow if you've followed this guide precisely.

Happy growing...

www.ingramcontent.com/pod-product-compliance
Lightning Source LLC
Chambersburg PA
CBHW031929240526
45464CB00023B/2812